I0109206

What Others are Saying About
Blood / Not Blood ... Then the Gates

"Ron Riekki stares into the violence and aftermath of prisons and poverty and addiction and especially war, sometimes unflinchingly, other times with such true, human flinches that I feel them in my own hands as I hold the pages. Individually, many of these poems are almost unbearable in their relentless and ugly truths, but taken collectively their beauty is that of a path, like Dante's, through what infernos we have made, to a saved life and—somehow—deeper humanity."

—Jonathan Johnson,
author of *May is an Island* and *The Desk on the Sea*

"Riekki does not leave us hopeless as he progresses into post-traumatic growth—'controlling my breath, my thoughts, my history that I was creating [...]' If you are sincere in your desire to support and understand Veterans, then you need to be reading contemporary Veteran literature."

—Suzanne S. Rancourt,
author of *murmurs at the gate* and *Billboards in the Clouds*

"Ron Riekki won't let you think poetry is just for your head. His book drags our bodies into the game. The work swirls and revisits past events, the visceral drumbeat of painful memories. The faint hope at the end gives perspective to where Riekki has been. This collection is like his description of swimming in a frigid river: *It's a sort of cure.*"

—Eric Chandler, author of *Kekekabic* and *Hugging This Rock*

"'The war is language,' wrote Allen Ginsberg in 'Wichita Vortext Sutra.' Ron Riekki's gone to war and proved that the pen is mightier than the M60 machine gun. In these poems, he douses flaming helicopters with the water of poetry, he translates tortured screams into English, into poetry, into sighs of relief. War is Hell, and Ron Riekki is the Virgil we need to walk us through it."

—Tom C. Hunley, author of *What Feels Like Love*

Other 21st Century War Poets
from Middle West Press

anthology

Our Best War Stories:
Prize-winning Poetry & Prose
from the Col. Darron L. Wright Memorial Awards
Edited by Christopher Lyke

❖　❖　❖

poetry collections

Always Ready: Poems from a Life
in the U.S. Coast Guard
by Benjamin B. White

September Eleventh: an epic poem, in fragments
by Amalie Flynn

Heat + Pressure
by Ben Weakley

Permanent Change of Station and *FORCES*
by Lisa Stice

Hugging This Rock:
Poems of Earth & Sky, Love & War
by Eric Chandler

Welcome to FOB Haiku:
War Poems from Inside the Wire
by Randy Brown, a.k.a. "Charlie Sherpa"

Blood /
Not Blood ...
Then the Gates

Poems

Ron Riekki
Middle West Press LLC
Johnston, Iowa

Blood / Not Blood
Copyright © 2022 by Ron Riekki

All rights reserved. Except for brief quotations in critical articles or reviews,
no part of this book may be reproduced
without prior written permission from the author or publisher.

❖   ❖   ❖

Poetry / Military Life / Trauma & Healing

Blood / Not Blood ... Then the Gates
by Ron Riekki

ISBN (print): 978-1-953665-12-6
ISBN (e-book): 978-1-953665-13-3
Library of Congress Control Number: 2022942151

❖   ❖   ❖

Middle West Press LLC
P.O. Box 1153
Johnston, Iowa 50131-9420
www.middlewestpress.com

❖   ❖   ❖

Special thanks to James Burns of Denver, Colorado!
Your patronage helps publish great military-themed writing!
www.aimingcircle.com

CONTENTS

After I got out of the military

I was still in the military
because of my body.

How do you not write about the body?

for Leah Poole Osowski,
Ellen Kane,
and Bob Flanagan

when the body is a switchblade, so far from switchgrass,
when the body seizures your attention and your underbelly isn't
and the body is the thing that allows you to describe the thing, but only
when the thing
is quiet
and my body is a backpack, a lost backpack, found stranded in an
airport
then my body turns to violence on itself, an anaphylactic heart-range,
hands fibrillating,
then my body becomes essential, but for its tremors, the doc says are
mental, psychogenic,
but that's mental
because my body decodes and derides me, decides to erode me, un-cliffs
and re-cliffs me,
handcuffs and regrets me
because my body has been thunderstormed by jobs, strawberried by
work, exposed to radiation, and worse,
when my body loves my soul, yet the world ate my body
and this body is almost ghost, and the world is hungry for everyone

The Marathon of Luck of Getting a Good PTSD Counselor at the VA

When you figure out how to inoculate intrusions,
the gift is sleep.
It takes one extra call, but there is such a thing as
a Mozart of psychology,
the way lives can not only be saved, but comet-like

strengthened,
the ability to not only breathe, but to clear out
the smoke
of panic. Death leaves footprints on the mind,
the crime scene

of war, how blood never washes out of memory,
but there are those
who help to build new moments, a family,
the gifts of birth
and rebirth.

RON RIEKKI

Watching the Search-and-Rescue Boats from the Shore During Desert Storm

Horizon   shifts meaning   boats scattered   the sun drowned

The Bus to MEPS

Those heading to college
will never know
the sacrifice
of even just
this bus ride,
the way that we
dedicate our lives
forever to this road
that will take us through
our lives—always a vet.

Sonnet #91: New Research Links Iraq Dust to Ill Soldiers

The doctor says *coffin* begins with *cough*, laughs.
I want to turn him into a panda, something on the edge
of extinction. Once, in third grade, I got in a fistfight
with a fist, lost. The doc tells me I have *pleural
apical scarring*, which roughly translates as "wood chest."
I steal a hair dryer for my girlfriend, the next three weeks
filled with the terror that the cops are going to come
crashing through my mirror, right when I'm trying to take
a leak or give a leak or get rid of a leak, especially
the one that keeps coming from the upstairs neighbor,
the apartment filled with spores. My girlfriend
is from Beijing. She says in America we're lucky
to be able to breathe. In China, air is an anaconda.
Eventually, everyone is going to have to work as a carnie.

PTSD

Whoever murders Santa Claus is going to be so famous
it's not funny. I mean, really not funny. It'll be sad,
especially when they realize it's just some minimum wage

senior citizen with a hacksaw face. We should make it
so that the news only has lit light bulbs. No close-ups
on shadows. The kids can't tell the difference between

video games and movies. OK, I've seen a helicopter
on fire. Let's get that out of the way. Let's put that image
back in the back of my head, the area that's protected

by thick skull, parasagittal posterior. The freak accident
in California. Skaggs Island. A base closed by Clinton.
We were playing baseball. This helicopter just falls

out of the sky. That's how violence happens. It tends
to have an impulse disorder. The guy who shot Lennon.
The idiots who eat St. Nicholas. The ones who crucify.

Sonnet #0: My PTSD Clings to the Center of My Christmas

like homeless children in homes, tachycardic
from the insecurity of walls, the way the devil
digs into your pulse, proves that escape is not
history, that the heat of the hole of your head
supplies you with a constant need for intrusions,
the wish the helicopter on fire in your youth
could be drowned in rain, the melted flesh inside
melting away with your patriotism, your black-
and-white photographs of death vermonted
to the high-five days of transcendence when God
existed as heavy as hate and now after the waiting
room is gone, after the counselor's shoes are in
the past, you almost see a dark bird of peace
approaching the holiday's police siren lights

The One-Time Return of Night Terrors

I beg the counselor
to help me return
to avoidance, but he
says I need to be
out in the crimson
pool of people,
that things get worse
before they get breathing
and I open my eyes
in the rivered room,
its throat of night,
and beg myself
to leave the doom-
birthdays of boot
camp and realize
fully that family
exists in rooms
nearby where
the father
cannot allow
himself
to be
strangled

I turn on
the light,
kneel
and insist
God
enter
every

9

RON RIEKKI

cloud
of me

insist

insist

insist

My PTSD poem

has PTSD. The poem itself gets triggered
simply by the word "trigger." Even the horse

makes me hoarse with worry, lose my voice,
makes me feel like life is a mixture of ice

and ICE, of bomb squads and vice squads
and when I was in the military, they put me

on the bomb squad, the pre-bomb squad,
the squad that had to look for UXOs before

the actual bomb squad would arrive, meaning
we'd walk slowly, so slowly, in a line rocking

our eyes from side to side, examining
the ground like our lives depended on it

and our lives depended on it and this poem
remembers and reading this poem reminds me

of how my eyes moved, as if reading this poem
is like scanning the ground, and there was one time

when I pissed my pants, not because of fear,
but because we were in MOPP gear and we

couldn't take it off and we couldn't go back
to our tent, but had to keep looking for what

ended up being nothing, nothing but this warm

RON RIEKKI

feeling on my leg that turned cold and reminded

me of the river in my hometown, the one that
the mines turned to muck, iron ore muck, how

I looked down from the bridge and saw the water
had died, except it kept moving, just like me.

The last word in PTSD is Disorder, which is not exactly the best word choice,

as if it's all about organization.
Although maybe it is. The organization
I was involved with was the military
and I go to the VA and in the waiting
room we wait, hard-core wait,
where it's hours and hours where I have time
to count all of the missing arms and legs,
the missing organs in these men
who joined the military to get money for college,
except they didn't realize how hard it is
to get through college with a traumatic brain injury
and I sit there on the hard chair
and think about "stress order,"
what it would be like to put all my stress
into a single-file line,
the way we'd stand in Boot Camp,
a camp for boots,
where I still can't believe
how much time we spent
in boots,
shining our boots,
and our drill instructor, swear to God, even calling us "boots,"
his nickname for us,
this man whose job was titled "drill instructor,"
whose job was to drill,
to make holes
and they did,
ten holes,
as ten people died

and I fall asleep and think of the yard of graves,
falling asleep,
falling,
fall in,
fell,
hell.

The Vet at *The Moth* Tells the Audience about his PTSD

He says he lost the woman of his life,
not his ex-wife,
another woman;
it gets a laugh.

It's his only laugh.
He says that he keeps having flashbacks

to a time when they had to pick up the bodies
after a firefight, a Taliban member
whose brain matter leaked all over his uniform.
He tells us that he recently picked up a gun

and put it in his mouth
and then threw it across the room

and decided from that point on
that he'd never kill himself.

I look at the audience
and they're in a church
that hurts—
America.

RON RIEKKI

Counting the Missing Arms and Legs
at the VA Medical Center, Orlando, Florida

and Chicago and Detroit and Los Angeles,
the ricocheting around so common to veterans,
how the military teaches you the nomad life,

gives you skills that so often don't apply
to the civilian world, despite what they say;
you don't hear them counting out loud,

all of the missing arms and legs that seem
to keep me awake night after night after summer;
afterwards it's like there's nothing left to say.

Looking in the Mirror During Desert Storm

You see yourself in yourself,
the murmuring incarceration

of B-52s. The base is a base;
it's the edge of something.

It's a thing. In the background
in the shadows in the barracks

in our room is my bunkmate,
a bulimic whose sleep is as

brutal as God. I watch him
watch nothing and then stare

into my guts, on the ledge
of the world.

My Bunkmates, Five Different Duty Stations

One did translations in Panama, told me of borderline torture where
he'd tell me how hard it was to understand screamed Spanish;

another got caught buying a hooker who was really a cop in Pensacola,
but they still let him try out for the SEALs;

another cried when I choked him in his sleep, but I got sick of his
making fun of me and telling me how rich his parents were;

another had the most beautiful black girl coming to our door everyday
asking if he was there, but his girlfriend back home got hit by a truck so
he'd wave that he wasn't in;

another used to do LSD, but said that now he was doing Buddhism and
it was all about staying clean;

another used to wear sunglasses everywhere and crank Billy Idol from
his car that I don't know how he was able to afford;

another liked to have a Confederate flag in his room until the
command told him to take it out and about ten months later his car got
burned down outside the base;

another wouldn't talk to me so I know nothing about him;

another had two crates full of porn that he would ship to every base
where he was stationed;

another was a white guy who got raised by a black family and he told me
that people always think he's black when they hear him on the phone;

another had the person next to him killed by small arms fire so that there was flecks of his buddy in his hair, in his eyes, on his neck, in his nose, in his mind, in his future

I tell My PTSD counselor
that I spent a hundred bucks

at a strip club
and I feel like hell.
He asks, "Why hell?"
And I say that I was suicidal.
And he says, "And it helped you forget?"
And I say, "Yeah."
And he says, "So?"
And I stare at a fish tank that has no fish in it.
It's just a tank.
And it's quiet.
And then the counselor tells me that soldiers used to bleed to death.
He says they'd get shot in the arm, the leg, and the medics would try to
save them.
They didn't realize the importance of tourniquets,
that you can lose the arm and save the life.
He asks me if I'd rather lose my arm or stay alive.
I tell him I'd rather stay alive.
There are a lot of tourniquets, he says,
just don't use them too often or you'll lose your arms, your legs.

(22 vets) PTSD stands for

I'm an *I am*.
I get shocks through my body
when I think of things.
The DSM tells me about frequent nightmares.
My bunkmate told me that the hard thing about translating the words
of someone being tortured is that you can't tell what the words are
when they're screaming.
The fist fights in high school were different from the fist fights in
middle school.
I know one person who died of leukemia,
one person who died of cancer,
and six people who died of suicide.
At the VA, after they took my X-ray, by mistake, they took the exact
same X-ray a second time.
It's not good to get an X-ray after an X-ray.
I had a job where the boss, as a joke, had me walk into an area that he
knew was filled with radiation.
It was a joke.
The last time I cried, it felt good.
The time before that, it felt like hell.
I do not know how hell feels, what it is like to touch hell.
Or maybe I do.
There was a syringe on the ground by the merry-go-round.
It is called 'avoidance symptoms.'
As I get older, I keep looking at my face.
What's happening to my face?

When the drill instructor would choke us

we wouldn't do anything.

I remember marching to the mess hall and seeing a raccoon that was
limping.
In the COVID wards, I remember all the stuffed animals
left on the beds
of the patients who'd died.
I asked the nurse what they do with them
and she shrugged.
The girl in class who said she's an alcoholic and she wants to help others
struggling with alcohol,
I wanted to go up to her after class and tell her
she's a hero,
but I was too afraid.

She wouldn't have been afraid.

PTSD Headaches

One Thanksgiving, I said to my mother,
"Have you ever seen bodies melt? No?
then give me a break." She broke, closed
her bedroom door, didn't come back out

until New Year's. I was in two services,
my brain cut in half. *Thank you for your
nervous disorder.* I do not stand during
the Pledge of Allegiance, because I can't.

RON RIEKKI

after I saw the people melting

in the helicopter on fire,
I can't remember anything
after that.
I also sometimes
watch hypnosis

on-line
where they hypnotize you
to forget.
My PTSD counselor told me
to be careful with that.

She said, *Be careful.*
I wasn't careful.
I kept doing it.
I wiped out another part of my mind
by punching myself in the skull.

I would do that over and over.
Over and over.
I had a seizure once.
I stopped.
I talked to the physician's assistant in the ER.

He said concussions are very dangerous,
that chronic concussions are even worse.
I said I didn't know what to do.
He just stared at me,
didn't say anything.

An Old Navy Buddy is Dead of COVID

He was a good guy. I don't know if he was
a good cop. Is there such a thing? I look
at the crowd at the funeral. It's raining.

It's cop rain. We all live too far north. Why
would anyone live up here? All there is
is wolves and howling and rat-like clouds.

RON RIEKKI

When I was in A-School, they drowned Lee,

another recruit,
and we weren't even at our first duty-station yet
and already one of us dead.

In football in eighth grade,
a kid broke his leg on the first day of practice.

In fiction workshop, the teacher called it *heightening*.

My Bunkmates in the Navy

One would meditate to death metal, center of the room, a non-vegan, the most carnivore I've known, how the years ate his intestines, seeing him after jail, the joy gone, replaced with bills.

Another who told me the bulimia was killing his teeth, murdering him, how I only heard him once, never knew, only saw his teeth when he cat-walked them for me, a Tuesday, a holiday.

Another slept with so many women he lost count—how? we were on an island, ninety-nine percent male, how they'd come knocking at the door and I'd be stunned, the ghosts alive.

Another carried a crate of porn with him wherever he went, had it shipped, this pirate's chest filled with bodies, all of these corpses he saved, stowed, never looked at, at least not with eyes.

Another who was an ex-pimp, told me, drank when we played basketball, said he was only good at sports when he was high and when he was sober he wouldn't speak, just threatening the moon.

And another who said he was going to be a professional golfer when he was free of his contract, telling us his father was a doctor, how he did this for him—what?—agreed to be trapped with us.

RON RIEKKI

Oh Christ, Not Another Poem about the Helicopter You Saw on Fire in the Navy

My God, haven't we already hit this hard enough, the flames,
yeah yeah yeah, the flames that looked like leaves, like gray
loves, like the history of guillotined guests, *yeah yeah yeah,*
we've all seen violence, yours was in the military. *Get over it,*
my PTSD counselor says, tells me to stop writing about it after
he told me to write about it till I get sick of it. *I'm sick of it,*
I tell him and he says good and he has a Marine haircut and he
was never a Marine and in the waiting room he catches me
reading a magazine about soldiers and he tells me to come in
and we walk down the hall and he says, *You know you're not
in the military anymore*, and I want to punch him in the fist,
tell him I know, but they force me to come here, on base,
for the counseling where the magazines in the waiting room

are all backward guns and camouflaged teeth, and we get
to his office and we sit there and sit there as we do, just
sitting there, and then he says, *so*, and I nod and we don't
talk about much of anything for the next forty-five minutes,
because I was late, missed the first fifteen, and he says if I
really care about this, I wouldn't be late and I don't tell him
on Veterans Day I punched myself in the parietal lobe six times
because I was trying to erase a memory and it ended up causing
a seizure and then caused something that I found on-line
that's called essential tremors where I earthquake myself to
non-sleep, the way that my mother tells me, *You should just
enjoy life*, and I tell her, *I'm worried I have brain damage*,
and she *tsks* her tongue and says, *Oh, how you exaggerate.*

When I was in A-School in the Navy

an instructor killed
one of the recruits
and only got eight months in prison.

In Boot Camp, I remember when the drill instructor
punched the kid next to me in the face.
That was the same day we were in blood—

I mean, the same day we were in line
to get our blood drawn. I'm shaking.
A Marine told me his PTSD

is like Jesus in the desert, Satan approaching,
and that time with Satan
goes on forever

In the Marines, I Had a Top-Secret Security Clearance

One of my coworkers told me he used to jack cars.
What do you mean? "I used to be a carjacker."
He had a Top-Secret clearance too. We got encrypted
messages that meant nothing to us. You could read them,
but why? It was just a vomit of the alphabet. A shitload
of x's. Every night, they'd have us buff the hallway floor.

We'd go out there and rap to each other, making up lyrics.
It was 3 a.m. ... 4 a.m. ... 5. The high of sleep deprivation.
This was during the war. Not a war you knew about.
It's strange knowing we were at war with a country
and when you come home you have to pretend like
it never happened. When I got the job in the prison,

I was surprised at all of the stabbings. I thought it'd be
prisoners stabbing each other, but they'd stab themselves
to get a free trip to the hospital, to get out of the prison.
In the military, we had three people commit suicide.
It got them out of the military. I know. I thought about it
too. I remember one night when I picked out all the clothes

I thought I'd be found in. It was a tough decision, knowing
what shirt you wanted to die in. But I never did it. Instead,
that night, before the cafeteria closed, I snuck in and stole
ice cream out of the officers' side. They got ice cream.
We didn't. Except I did, that night. It felt nice,
being alive and eating stolen Rocky Road.

Night Terrors

Nightmares sometimes
where I'm being tortured.
In the military

my bunkmate
was a translator,
told me how hard it is

to translate
when they're screaming.
My uncle,

a heroin addict,
woke up one time
inside a body bag.

The medics helped him
get back out.
I've seen a helicopter on fire;

my PTSD counselor told me
to write about it so many times
that I'd get sick of it.

The panic attacks tripled.
He said, "Let's have you
stop writing about it

for a while. OK?"
I'm not supposed to be writing
this.

Sonnet #13: Blood, the Blood, Not Blood, Then the Gates

I have no idea how you make money without arteries.
In the military, we had a guy who—
I won't tell you. Like you need all the bodies.
At least one guy died on every base where I was stationed.
You'd look up and see a cornfield; you'd look down
and see a bloodfield, a veinfield, a windfield, a scarfield,
the shadows would eat the shade. You'd get killed
by shrapnel and friendly suicide and fire
that would glow like the gloves in your gore.
In the military, we had a guy who put a rifle in his mouth
on qual day, showing marksmanship with mist like wine;
I remember the instructor doing chest compressions
on a guy with no head. When the haunting starts
everybody pretends there's no ghosts.

OK, This One's About the Suicides in the Army

Actually, the first guy was a grunt. This was Boot Camp.
It wasn't my Boot Camp. It was two days before I got there.
The first guy was a groan. They say that the soul sticks around

for a few days after the shame. God, death looks so boring.
It's like mathematics gone bad, dirty science.
I know a chick who works as a mortician and she said she gets sick

of being awake. The corpses, she said, get their beauty sleep
forever. And then she laughed like a duck. It's what you have to do,
get down before a bullet ends up in your church.

Another guy drank himself to death.
    This was in the Brit Marines barracks.
Those guys had no trouble killing hallelujahs. For their PT they'd jog
with the other guy on their back, switch, and keep going.
    It reminded me

of Jesus, except with the F-word: *And know that I am with you always;*
yes, to the end of fucking time. Another jumped off a seventh floor
in Spain, drunk off the moon. He drank the guts out of it.

You can drink outskirts, eat battalions, suck aftermath; my canteen
made me puke. "Suicide sweeps California Marine Corps Unit."
'Sweeps'?! A buddy said, "There are absolutely no benefits
    to being a vet,

but there are goddamn tornadoes' worth of bad shit that comes with it."
He punched a stop sign, as if to show how much a fist can keep going.
    Sweeps?
I'll give you motherfuckin' "sweeps" ... The second guy was a scream.

33

RON RIEKKI

They say that

dedicated to Negaunee

one out of every two Americans
suffers from PTSD,
that the country
is a body,
the rigor mortis
of Texas.
In elementary school,
I remember walking home
and seeing three boys
killing a cat
with sticks,
like it was some pagan ritual
where the sons of miners—

with their fathers caked down
into the earth,
absent and angry—
were trying to beat the innocent
so deep into the ground
that God would emerge
and choke out
all the poverty
of existing
to quiet them.
I kept walking;
like God,
did nothing.

Reading an Article about *American Sniper* at the VA Medical Center

for Leesburg

The doctor says, *You know, you're no longer
in the military.* Yes, I know, but my lungs
are still back there, so I can read whatever

the hell I want. That's gangsta—the guy
with no legs next to the guy with no arms,
both of them hungry as hell waiting for test

results to come back and let them know
their legs and arms are still back there,
so they can read whatever the fuck they want.

RON RIEKKI

I Sometimes Write about Violence

have found an audience
for my violent poems.

I was in the military,
served during Desert Storm.

I worked in a prison,
got attacked by an inmate.

(I probably would have attacked
me too, if I was doing life there.)

I worked as an EMT
in Los Angeles.

I tell a friend of mine,
a gay Brazilian poet,

about my violent poems
and he tells me to stop writing

violent poems. I ask him
what I should write about.

He says to write about praying
or lying in the grass in the field

across from his house
or my house, any grass I can find.

Or, he says, take a step back.

Write about you writing about

your violence. Write about how
you publish your violent poems,

which encourages you to write
more violent poems to please your

audience that wants your violent poems.
And slowly start to weave in

grass and love and the moments
where there is peace. There are so many

moments of peace, he says.
Like now.

In the military, they made me go around the building, killing birds

I swung the broom like a child. I put on a show,
as if I was really trying to kill them, but I wasn't.
I couldn't kill anyone. I don't know what I was
doing in the military. I couldn't harm a fly.

Actually, I should be truthful. I've killed a lot
of flies. They're hard to kill too, so maybe not
that many. But some. I remember when I worked
in the prison, there was this fly strip, all these flies

glued to it, about fifty flies, the whole thing full.
There was barely any room for any more flies to stick.
They never took it down. I was there for years.
That same fly strip. Always the same. So useless.

But always there. On the day I knew I was quitting,
I thought of taking it down. What would I do with it?
There were no garbage cans in there. There was nothing
in the prison. Just these inmates, barely any room for them.

(PTSD) Sometimes

I go to the river.
Even in the winter.

Especially in the winter.
Whenever I get triggered.

My body quivers.
I can't get it to stop.

I'll walk down to the water.
I'll take off my top, my socks,

shoes. I'll toss them on the shore.
So quietly, horizontally, the snow

falling. And I'll hover
there for a moment, then, slowly,

get in. *The sad effect of sadder
groans*—a George Herbert

line I like, a poem called "Death." So
many good poems I'll never read. I saw

a man in a house one time, some
retired soul, watching my lack of sanity,

working my way to the center,
midway between shorelines, where

I'd lose any sense of feeling in my fingers,
my legs, my thoughts. It's a sort of cure.

In the military, they made us paint the bottom of stairs.

They made us wash the backs of the tombstones.
We had to wash the dishes that no one had used that day.
We had to replace any oxygen we'd used in the barracks,
go outside and capture it in buckets, bring it back, put it
in the exact spot where we'd breathed it originally.

We'd shine our shoes, shine our pillows, shine our pneumonias.
Everyone got pneumonia in Boot Camp. It's what Boot Camp is for.
It's also for boots. We shined the hell out of those boots.
It's not really a camp though. It's more about the boots.
The socks meant nothing. The easiest thing in the world to fold.

And there was so much folding in Boot Camp. We folded
our underwear and folded our pants and folded our letters
and folded our lives and folded our futures and folded the walls
with our fevers so that the walls teared away and we could see God
staring down at us in our sleepless nights where we all coughed

in unison, aching to go home, but there was no home anymore.
Instead there were all these stairs to paint. But only the bottoms.
So that the paint could drip into our faces. With each drop of paint
that lands in your eyes, it's one more step of having your childhood
erased.

It feels wonderful when a war begins, they tell us. And then a war
begins.

I loved thunder as a child.

And ghosts. I wanted there to be
more ghosts in the world. My father
told me that probably about a hundred
billion people have died in the history
of history. That means a hundred billion
ghosts, he said. I think he was trying
to cheer me up. Where are they? I asked.
He said they hide. A hundred billion of them?
Where would they hide? Under your bed,
my father said. He was trying to cheer me up.

I always thought there was one monster
under each bed, but maybe he's right.
Maybe it's a billion of them. I waited
until it was a thunderous night, kept
the light off, got down and looked under
my bed. And there they were. A hundred
billion ghosts. All of the dead from the entire
history of the world, under my bed.
It looked like nothing, but I knew they were there.
Two-hundred billion absent eyes staring at us.

Ten People Died When I Was in the Military

I worked in a prison.
The psych ward of a prison.
It was a prison
where they sent the most dangerous prisoners
from all of the other prisons

in the state.
I hated the state
where I lived. It was a southern state
that thought it was the best state
in the best country in the best world. It was the state

that spent the least on prisoners
out of any prison

in the entire country. They were so proud of that.

(vets per day)

I hate the voice of heroin, calling a relative to say hi
and heroin's there, in the voice, how just telling them
that I'm trying to say hello becomes a long, slow
conversation of them saying they don't understand
and I tell them I'm worried about them and my uncle,
for example, responds, *Don't worry about me, Ronny,*
except taking a full minute to say five words, possibility
of overdose, which has happened before, a friend bragging
about being dead, which is like bragging about your
ability to brag, or like dragging your corpse across
another corpse, or worse, or what's worse? and then they
tell me, how heroin is horror, how it's a genre, injecting
into places you don't want to imagine, in places you
don't imagine, and it's the absence of imagination, while
I watch one relative's face that looked like Jesus before
and looks like skull now, the skull of Jesus, the skin
gone, the eye sockets as light sockets and an inability
to remember my mother's name, and inability is really
what heroin is, an invisibility, the eraser of the good
for, no, not the evil, for the neutral, or for the absence
of feeling, or no, there is feeling, plenty of hurt, but
really what it is, I think, is sheer absence. Just absence.

RON RIEKKI

in the military

they take your back. They don't give
it back. They don't give a shit. They

take you to the camp where they give
you boots, kick you in the teeth. I had

no idea they could hit us, whispered that to
another recruit, and he paused, said back, *You
do realize we're about to go to war, right?*

Mankind

"... one less ..."
— Robert Frost

I've always worked with violence, escorted
violence, introduced violence to violence
and told violence to let me know if violence
got out of hand. An elementary school
violence that turned into middle school
violence that turned into a high school
violence that, truth be told, reminded me,
years later, of a gull pecking at a cat corpse
on the other side of the prison fence, but
I'm moving too fast, because first there is
the military violence that PTSDs you and
gets you ready to be sent back to the street
violence, except this time with so many

concussions that the brain has lost count
and lost other things too, and that gull
with its wings that I wanted to pluck and
insert into the backs of every inmate I saw
in that place, how I was incarcerated for
thirteen hours at a time, placed in nursing
stations with shatterproof glass that one
night was shattered, an inmate kicking it
in until right behind him thirteen guards
lined up in their cockroachish riot gear,
all shadow and threat, and there was this
moment of peace, a strange moment of
peace, and then they all, in synch, attacked

The social worker interviewing me frowns when I say

"[...] Back then, I wasn't shit. / Just electrified violence. [...]"
—Rachel McKibbens, "one more time, with feeling"

'veterans.' Apparently, this is the wrong answer.
The question: who I want to work with. And I work
with veterans now. Is he frowning at that too?
And why the frown? Is it genetics? Sunlight forcing
the edges of his mouth down? Or a hatred of veterans?
No one hates veterans, my mother'd said. And then I
filled her in. The question I hate the most: *Have you ever
killed anyone?* I've been asked this in banks, on dates,
in bed. I've been asked this on the beach, on my porch,
in Detroit, Michigan. I usually answer with nothing, my
body too shocked to speak. And then time passes, some-
times decades. And then I'm in front of this social worker
again and he fills in the silence with another question.

This time he wants to know how many people I've killed,
except he's asked me something else, but that's what I hear
and I'm gone and the job's gone and the moment's gone
and I remember to breathe. Even after Zoom has died
and gone back to the hells of cyberspace. And I breathe
in my room. It's what I've been taught. And then I get
hired somewhere else. I teach veterans how to breathe.
I learn that twenty-two veterans per day kill themselves.
And I have one living, in front of me, now, and I teach him
that twenty-one percent of air is oxygen, how beautiful it is
to breathe oxygen, and I teach him what I was taught, which
is how to breathe it all in, complete presence in this moment,
and then I watch him breathing, love that he's breathing ...

The dead from when I was in the military

"... one less ..."
—Rynn Williams, "Reflections in Porcelain"

bore those who were not
in the military, but
propose to me at the side
of my bed, all ten
of them, all wearing
nothing, because they
are dead, no skin; I turn
on the light, quick, to
make them all even
more dead, but there
they are, all on one
knee, asking me to re-
member forever ...

RON RIEKKI

PTSD

I lock the doors. All
the doors. I lock
the windows. All
the windows. I lock
the lamps. Even
the lamps that are
in other apartments.
They ask me what
I'm doing there.
I tell them it's safer
when the mountains'
cinematography
is lit for all to see.

22

We're all waiting. It is a waiting room.
Except there has been no thought about people.
It would be perfect if it was never used.
It's the type of room that looks like it should never be used.
But it's used.
It's packed.
Chairs that hurt.
This is the VA.
Here for post-traumatic stress disorder counseling.
Many of us are triggered by such things as, well, triggers.
Things like seeing guns.
And when we walk into the front door to the building,
there is a man with a gun.
Security.
He looks pissed.
Always.
Every time I have ever seen him.
Because when you are going for counseling, it's important that you are
immediately greeted with anger.
And the person who checks us in
mispronounces my name
every time.
Every.
I correct her.
Him.
There are two front-desk workers.
Neither says my name right.
Ever.
Even after I tell them.
Multiple times.
I can't imagine if I gave them the pronouns I prefer.

It would blow their minds.
They would combust.
I go for counseling.
The counselor has never been in the military.
Maybe not necessary.
But maybe it is.
I don't know.
My counselor tells me he's thinking of quitting.
And then we start the session.

WHOLE (ptsd)

"... will be 'safe' ..."
—Alan Pelaez Lopez,
"Zapotec Crossers (or, Haiku I Write Post-PTSD Nightmares)"

i.

One counselor tells me not to write about it.
Tells me to write about everything else.
Tells me to write about gratitude.

The things I love.
I love to write.
I wonder if I love to write about PTSD.

It gives me control.
I can turn my PTSD into a fish.
I can drown it in water.

I can drown it in the sky.
I can have my fish fall into the whole of the cosmos,
how deep the sky is,

this fish that is, now, my PTSD dropping upwards
forever
until it cannot harm me anymore.

ii.

Another counselor tells me to write about it nonstop,
tells me to take the helicopter on fire
and the plane crashed into the ocean

and the gun-in-mouth Marine and the
drank-himself-to-death Marine
and the killed-by-the-Guardia-Civil seaman on my shift

and the drowned-by-the-instructor-during-training kid (and he was a
kid, from the Midwest, like me)
and to write them

over and over
until they stop affecting me,
until they get sick of standing around on the page,

get sick of standing around my bed at night,
get sick of being in the same helicopter on fire
forever

and instead
go
to the land of the dead

and how you can drain yourself
of memory
by getting sick of memory

and the counselor tells me
whispers
write in third person

so that I
lose the I
this deep sense of connection to my self

and I become
Ron Riekki
this thing

this character
how Ron Riekki has PTSD
but Ron Riekki is not real

is a memory
how the helicopter is not here anymore
how it's so far away

on the other side of the cosmos
how the memories
of our traumas

are just memories
and when we put them on the page
and distance ourselves from them

not recreate it in a lie where we pretend it's now
when it's not
but we make ourselves a character

who can do anything
Ron Riekki is a lion
Ron Riekki is a hero

Ron Riekki can walk on water
and walk under water

RON RIEKKI

and walk without water

and can become water
all of the water in the world
Ron Riekki is an ocean

and she said
that if I keep writing
and keep controlling

the narrative
I will
one day

free myself
from the memory
because I will become the sky

I'm ...

getting a degree
in Substance Abuse Disorders
just one degree
just that little bit of heat
where you take something
and give it a slight fever
where history—
yes, God, goddess, my goodness, that brutal tsunami
called history—
is (let's run through this quick)
my father, age, oh, eight, came home from playing baseball with
kids in Palmer came home, found his mom dead, she'd drank [drunk?]
herself to death, he didn't understand death, dragged her from the
kitchen to her bed, tucked her in, and, of course, you know the rest,
how it's silence, how you can't share after that, turn off, and, well, my
uncle overdosed on heroin, got put in a body bag by the medics, except
he wasn't dead, started punching to get out, no zippers on the inside,
and the shock of them realizing, and then how I'd punch myself in the
head over and over and over and over to stop myself from using, try to
knock myself out, succeed, sometimes, rupture, blast, thinned,
succeeding, lying on the ground, not wanting to use because now I was
not wanting to move, chronic concussions, and so I shake, asking the
doc, is it Parkinson's, he shrugs, they shrug a lot, idiopathic, how hard it
is to sleep now because of the constant tremors how I thought for the
longest, honestly, that I lived in an area where there were daily subtle
earthquakes, and a cousin, got drunk, noon, yes, noon, that's how bad it
was, and he drove, get this, into a church, yes, but it gets better, meaning
worse, where they were, at the time, having an AA meeting, and, do you
want me to keep going? I should stop here
I'm getting a degree in substance abuse disorders
my family isn't proud

they're *PROUD*

no

they're desperate for salvation

no

they're being choked to death
by history and that one solitary degree, that little bit of Fahrenheit will
hopefully burn the ticks so that they let go of the skin and the disease
will not have its chance to take over further like—and this is what it
reminds me of—the scene from *The Thing*, yes, that one, the worst one,
that's the abuse, because disorder is too soft of a word, substance, too
soft, abuse, yes, to burn the hell out of abuse and then time for time to
take its course and to let the soul heal and skin heal, childhood home as
burn unit

where the cure is stillness

and years

Even my PTSD has PTSD

and the VA waiting room is the color
of spit. And the drill instructor drilled
the most useless crap into our heads.
And by the time I got out, eight of us

were dead. And I'm a quote-unquote
'protected disabled vet.' Protected?
They could hit us back then. Dunno
if they still do it now. The doctor

called it "essential tremors." Essential?
My body earthquakes. My counselor
told me that the body keeps the score.
But the problem is my life hasn't been

sports; it's been horror. I need more
sex. But I'm too poor. At the VA
waiting room, waiting, I count all
the missing arms and legs. I get to four.

RON RIEKKI

At the dentist

when they grind down my tooth

I think of the crucifixions
in the military
where they'd come in the room—
you'd hear them exploding down the hall—
and they'd tie you to the chair
and then tie you to the fence
and stuff weeks-old food
in your mouth
for fun
to celebrate the boiling hot stupidity of
hazing

and

I'd grip the chair
like I was wanting to stone its children
to death
and the dentist
had no idea
how to deal with panic attacks
so she'd talk to me
about Snuffleupagus
which just made me beg for the ending
of all of humanity

in the military

when we killed people
and it was kept secret
so that it wasn't on TV
back home,
I'd sit there and stare
at the television,
how they'd show
David Letterman
at noon and then I'd
sleep the rest of
the day and go back
for the midnight shift
where the cemeteries
of the world
would all wait for us
in anticipation.

RON RIEKKI

in the military

one kid drank himself to death, a Brit Marine,
caution tape swinging like a ghost sitting on it,
and three more died in a plane crash and we

thought it was six, but four bodies were found,
three alive, and I stood on the shore and watched
the Search and Rescue the sea the moon the crabs

so many crabs the moon the crabs the sea the boats
how I went to the mirror in my barracks room
and punched myself in the skull, how I'd do that

after each death, three more in a helicopter crash
and the mirror my skull the night and them,
how we looked through binoculars, caught

in electrical wire, how the statistics seem off
if you ask me, my skull the bodies the barracks
the tape another killed by the Guardia Civil

thrown from a building, how they taught us
to run if the police ever came, you're in a foreign
country, they said, and we were and weren't,

we were on a base with beer-vending machines
and, God, there was all that porn in the bookstore,
not a single thing by Harper Lee or Orwell or

anyone, really, but the porn, and the beer,
twenty-four-seven, the beer, and then came
the suicides, two, one, an M16 in mouth, and

one, rope, and the mirror the tape binoculars
skulls of skulls, the cutting I started, taking
a paperclip and just rubbing it back and forth

on my ankle, which I thought no one would
see, the ease, the night, the day, the deaths
that were disappearances, so quiet, absence,

how one was married, and I saw his wife,
how they let her out. *They let her out.*
That's the term that was used.

RON RIEKKI

My PTSD counselor

tells me to take the flames and turn them
into roses, that I have the power to do that,
to control my own mind, that I can't control
any other minds, just mine, turning the red
of fire to the red of flowers and she tells me
to practice the next time the intrusions come
and they come, the triggers, so many, how
we labeled them, had me list them, notes,
any time they happened, and it happened,
in a garden, of all places, where I thought
for sure I'd be safe, my stomach turning,
how the counselor told me to write poems,
how the word verse comes from "to turn,"
how it comes from agriculture, how poems
are like rows in a garden, how we'd stand
at attention, all of us facing each other,
how I pissed myself once during inspection,
my pant leg turning a different color, the kid
across from me watching. We were children.

PTSD Paperwork

Statement in support of claim of
death. The downloadable pdf
is available for your headaches.
Related forms include suicidal
ideation and learning how to cope
with cutting. If you are not en-
rolled in a current divorce pro-
ceding, you can go directly to
the waiting room. Mental health
services include waiting and more
paperwork. When you are doing
the paperwork, you can work in
breaks of waiting and while you
are waiting, you will be able to
stare at other vets who are also
waiting. The waiting rooms will
be bland, but have strong hints
of passed traumas if you look
closely or smell weakly or think
unclearly or touch anything.

RON RIEKKI

In This Poem, I Talk about My Suicide Attempt

When a roommate asked me the difference between
Veterans Day and Memorial Day, I told him it was
the difference between waking up in a stranger's
bathtub and getting your stomach pumped. When
a girl I was dating asked me what a hurt locker was,

I walked up to one of the lockers at the gym and punched
it so hard that my fist opened up and you could see the knives
inside my wrists. When my ex-wife asked me why they call it
Boot Camp, I told her it was a cross between pulling yourself
up by your bootstraps and Camp Crystal Lake. I changed my mind,

said it was half S&M bondage shoe and half Belzec. I changed
my mind again, told her it's basically getting kicked over and over
again as a child. I showed her a video of the kid who killed himself
in the bunk next to mine, everything in slo-mo. I showed her a photo
of every service member killed in every war in the history of souls.

Then my roommate asked, "What's Boot Camp like?"

Go outside now—it's forty-eight degrees out—take your shirt off and start running around the apartment complex and don't stop until you can't breathe. What's war like? Go outside now—it's twenty-six degrees out—take all your clothes off and then start banging your head against the apartment complex walls and don't stop until you're insane. What's it like seeing your first corpse? Go outside— keep all your clothes on—and walk until you see a corpse.

The Helicopter on Fire
When I Was in the Military

The helicopter on fire when I was on the military.
The hell of fire when he was or the military.
The hell of hell when hell was in my military.
I'm hell because of the hell that was hell in the hell.
The hell I held helled and I fell into a view of the helicopter.
We looked through binoculars and watched the bodies melt
and I felt like looking through the binoculars and batching the belt.
The helicopter on fire feared the future of my history.
My traumatic brain injury was also in the military.
The burn bin that I waded in before where I breathed in the cancerous
skin.
The hell of cops, the killing fires, was I in the military?
I was and am and will be and was and fire is fire and no, no scenery.
The hell I say and say no more because the melting bodies melt for
eternity.

Thank you for your service, she said. (Said exceptionally flatly.)

. . . ‾ ‾ ‾ . . .   . . . ‾ ‾ ‾ . . .   . . . ‾ ‾ ‾ . . .   . . . ‾ ‾ ‾ . . .

Ode to Those Who Have No Real Comprehension of PTSD

I

The military, as we know, reduces
everything down to a few letters
and numbers, abbreviations and
statistics and when I look at the
statistics in the waiting room,

waiting to find out their latest
medical statistics, I realize that
the statistics are humans, that
stats breathe, that stats have
a pulse, or, at least these ones

do, as the dead are also stats,
have a breath of sorts, in me-
mory, the hyphens of death,
the hyphens of memory, how
we connect the stat with the

heart, the stat with the tomb-
stone, the way that 22 of us
kill themselves every day
and this beauty of 22 how
it repeats, how the statistic

is repeated, as if repetition
means anything, as if all
the pushups in Boot Camp
will save us from strange

RON RIEKKI

lung cancers where the doc

says, "I haven't seen this
one before." And he says
one, a number that reminds
me of *none*, how it's hidden
inside, swallows, the 0 as

a hole that we dove into,
hid, had our youths drain
away doing things like
painting the underside
of stairs and staring at

stars while B-52s started
their engines in the back-
ground, soon to change
the ground from even
to odd, soon to tear into

the earth, and our hearts'
stats come back—They
scream to us in the waiting
room, screaming out a number:
22. 22. 22. 22! 22!

Ode to Those Who Have
No Real Comprehension of PTSD
I I

Blood. A job posting saying *We're hiring*
vets and I notice it's a job that involves
blood. And I think why has *vet* become
synonymous with *blood*? Why aren't
the cush jobs for vets? Why does it have

to be the jobs where you can die? Is it
because we are expendable? A vet
I know says, why does it always have to
be unclear? I'm glad I'm seen as an anchor,
but I'd like a job that's not nuclear. If I say

anything, it's seen as rancor, as if I'm
a cancer. I want to tell them, no, I just have it.
I want to tell them, no, I could do hospitality
as well as hospital security, that I could
be a CEO and not just do voiceovers,

that I don't have to be hidden, jobs
where I stand at attention for ten hours
as if my feet are made of cement, plaster,
resin, mortar. Mortar. Mortar mortar mortar
mortar. Minimum wage rhymes with murder.

How tombstone and bomb are the same word.
Word. I wish society could be rewired,
so that we wouldn't get out of the military
and get sent back to a neighborhood

where we were poor before but now here's

some PTSD and what does the P stand for?
It all starts out with poverty. Ends with dis-
order. There is an order to oppression.
TS stands for thunderstorm, how much
lightning hits our bones, destroys our homes,

throws us down, and we have so much
patriotism that it riots in our insides,
in Detroit, in Fort Worth, in Portland,
in Arlington? What rots? The dead. We're
still alive. Give us a goddamn chance.

Ode to Those Who Have
No Real Comprehension of PTSD
III

Ten people died when I was in the service.
Ten people died and I'm nervous writing this,
nervous that you will not count the two

suicides, that they will be erased, that you will
only want to know the eight, and three
of them were civilians, so then it is only five

and one of them was killed during training
and do we count him? Can we count at all?
What do we count? How do we count?

Who counts? I can count. I can count
to a trillion. I'm sorry, but I can and
the three killed in the helicopter on fire

doesn't count because we can't look at it,
the red too bright, the flames too hot,
the war too fatherless, the night too dark,

the wind too still, the memory too haunted,
the food too dry, the milk too non-existent,
the ocean too long, the search-and-rescue

too distant, and there is one body left
and it comes down, falls from the building,
and do we hear it? Do we choose to?

RON RIEKKI

Ode to Those Who Have
No Real Comprehension of PTSD
I I I I

Is there comedy in PTSD? Ask the counselor
who doesn't understand why I couldn't touch
a computer for a decade, the way that we kill
people digitally, how we were at war with a
country that we weren't at war with and how
do I explain to the counselor that he doesn't
understand history because I was stuffed
inside history, puked into history, drilled
and killed and choked and bathed in piss's

history, the hoaxes of news, how it axes up
everything and turns it into entertainment
when there is nothing entertaining about
the leaking of brain fluid from the ear on-
to your uniform, their form lit by hell, un-
derstand starts with *un-* because there is

an undoing, an ungluing, how I honestly
felt my body coming apart, my insides
cracking, separating during the war, as

if my skin was sinking, as if the people
who died were taking atoms from my
spine and skull, and a girl I know had

her best friend explode next to her so

that little pieces of her buddy embedded
into the left side of her face, pieces of

his bone in her cheeks and the doctor
said, *I won't be able to take them all
out* and she said, *Good*.

RON RIEKKI

Ode to Those Who Have No Real Comprehension of PTSD
I I I I *I Exist*

I forget the names of the dead
sometimes and then make myself
look them up to remember the dead,

all of the dead that I know as if
their names are ghosts, as if even
the three words of what we pretended

made them who they were could just
fade in and out of the mind with the ease
of a wave, of the turn of a page, and

it's the opposite of their deaths, how
there is no violence (deceptively)
to the opening of a notebook or

to a simple google search where it
recounts the who, the what, the where,
the how, and the why completely ignored.

My PTSD counselor told me to

write this poem. She said for me to
capture gratitude on the page. No,
not capture, but caption, to enrapture

the good, to claim it. *But what?* i said
and she said for me to find that, search
for it, little kisses of bliss, of grace,

to not talk about the helicopter or the—
she didn't even want me to say the nouns
of trauma, to silence the shudders, to

shove the ghosts aside to see the living,
and so i drove home, crying, as i cry
after every session, tears like tides,

the rituals of the earth, how we repeat
for the good and bad—rain and tornado,
snow and hurricane ... and sun, soft

breeze—i stopped driving home. i
always drive home, barricade—embarrassed
to say this—my door, chairs blocked

up against it, it locked, but instead
i went to the beach, so nearby, not
where i live, but nearby, with effort,

crossing the bridge that introduces
the water, says, *yes, we are always
right here, whenever you need clean*

RON RIEKKI

air and so i, healthily, went, the good
choice, driving passed the prison, so
oddly placed, as all horrors are, and

i said a prayer silently for the bodies
inside, how they merge together,
their sickness and sadness becoming

one and i pleaded for my angel to
leave me for a moment and to go
to them and it did and i drove to

the sand where the goodness is so
obvious, and the day ebbed, and
i worked on controlling my breath,

my thoughts, my history, that i
was creating, and it seized me,
when I was there, seized me—

peace.

There is a Lull at the Dinner
after the Ceremony
Where the Vets All Sit in a Corner
Having Eaten Everything on Their Plates and
One of Them Puts His Spoon
Through the Ketchup Left on His Plate
and Says

I remember picking up a body and the brains leaking out
all over my pants

The Seconds When I Left
the Counseling Office and Felt the PTSD
Might Be Leaving This Time Forever

I'd arrived by foot, but the parking lot greeted me anyway with a gala
of sun, as if the past, the onomapoetic *plot!,* was barbecued, ready to be
eaten, shared, trashed; I wish this hilly bliss could blizzard the vets

drowning in the intrusive flashback nets, could fix their hashed exits
with the new days of *denouements*, to be rescued by relief,

decluttered, as if ungunned, the agony gone, fileted;

there are times you feel good enough to grace God with belief.

A Few Words of Thanks

I would like to dedicate *Blood / Not Blood ... Then the Gates* to Sally Brunk, my Aunt Jean, my Auntie Cookie, and my Uncle Paul, all of whom passed away this summer.

Further, I would like to thank Sally Brunk, Melinda Moustakis, April Lindala, Steve Balderson, Steve Benetier, Alex Vartan Gubbins, Aishu Giriraju, Danny Oseroff, Sharmila Voorakkara, John Bullock, Vivian Faith Prescott, Donald Hall, Randy Brown, Jan Schafer, Lauren Lowell, Ace Boggess, Joy Gaines-Friedler, Ben Weakley, Miskopwaaganikwe Leora Tadgerson, Lee, and my parents.

Thanks to the editors of literary journals who have featured my writing in their pages.

Finally, thanks to fellow poets Eric Chandler, Jonathan Johnson, Suzanne Rancourt, and Tom Hunley for their insights and their kind, pre-publication endorsements of this collection. The work of each of these creators connects to themes of land and love, and I am both glad and humbled to be in their company.

RON RIEKKI

Acknowledgements

I am grateful to the editors of the following magazines and journals, in which some of these poems first appeared—some in slightly different versions:

"The Bus to MEPS" first appeared in *O-Dark-Thirty* journal Vol. 6, No. 1, Fall 2017

"Watching the Search-and-Rescue Boats from the Shore During Desert Storm" first appeared in *O-Dark-Thirty* journal Vol. 6, No. 1, Fall 2017

"The Last Word in PTSD is Disorder, Which is Not Exactly the Best Word Choice" first appeared in *Nixes Mate Review* issue No. 12, Fall 2019

"The Marathon of Luck of Getting a Good PTSD Counselor at the VA" first appeared in *O-Dark-Thirty* Vol. 6, No. 1, Fall 2017

"My Bunkmates, Five Different Duty Stations" first appeared in *Stone Canoe* No. 11, Spring 2017

"My PTSD Poem" first appeared in *Jabberwock Review* Issue No. 41.1, Summer/Fall 2020

"Ode to Those Who Have No Real Comprehension of PTSD I" first appeared in *Collateral Journal* issue No. 3.2, May 2019

"Ode to Those Who Have No Real Comprehension of PTSD I I I" first appeared in *Collateral Journal* issue No. 3.2, May 2019

"Ode to Those Who Have No Real Comprehension of PTSD I I I I

I Exist" first appeared in *Collateral Journal* issue No. 3.2, May 2019

"The Seconds When I Left the Counseling Office and Felt the PTSD Might Be Leaving This Time Forever" first appeared in *Dunes Review* Vol. 20, No. 2, Fall 2016

"The One-Time Return of Night Terrors" first appeared *r.kv.r.y. journal* Vol. IX, No. 3 Summer 2017

"PTSD Paperwork" first appeared in *Collateral Journal* issue No. 6.1, Fall 2021

"Sonnet 0: My PTSD Clings to the Center of My Christmas" first appeared in *r.kv.r.y. journal* Vol. IX, No. 4, Fall 2017

"Sonnet 13: Blood, the Blood, Not Blood, Then the Gates" first appeared in *SAND Journal* issue No. 14, Summer 2017

"Sonnet 91: New Research Links Iraq Dust to Ill Soldiers" first appeared in *Little Patuxent Review*, Summer 2015

Glossary

A-School: After graduating from their initial Boot Camp experience, new recruits into the U.S. Navy move on to "Accession Training," where they receive technical instruction on their specific jobs.

B-52: A nuclear-capable, jet-powered strategic bomber manufactured by Boeing Corp. and first placed into U.S. Air Force service in 1955.

Belzec: A Nazi extermination camp located in German-occupied Poland during World War 2. Approximately 434,500 Jews, and an undetermined number of Poles and Roma, were murdered there between March and December 1942.

"Boot Camp": An informal term for initial recruit training. Geographic locations differ by military branch of service.

COVID: Coronavirus disease (COVID-19) is an infectious disease caused by the SARS-CoV-2 virus.

Desert Storm: Operation Desert Storm, sometimes referred to as the First Gulf War, is the American name for an armed air, land, and sea campaign involving a coalition of 35 countries, with an objective of ejecting Iraqi military from occupied Kuwait.

DSM: Regularly updated in numbered editions by the American Psychological Association, *The Diagnostic and Statistical Manual of Mental Disorders* provides a common classification system and language for medical practitioners.

ER: "Emergency Room."

Guardia Civil: The Civil Guard is one of two national law enforcement forces ("gendarmerie") in Spain, and operates under the authority of either/both the Ministry of Interior or Ministry of Defense. It includes air and naval functions. On peacekeeping and other missions, its personnel have participated in deployments to Iraq, Afghanistan, and other locations.

ICE: Immigration Customs Enforcement. An agency of the U.S. government.

LSD: A semi-synthetic organic illicit compound ("Lysergic acid Diethylamide," a.k.a. "acid") that causes extreme sensory distortions, altered perceptions of reality, and intense emotional states.

MEPS: "Military Entrance Processing Station." A place at which would-be military recruits are first processed for medical and administrative actions, prior to taking and oath of enlistment, and subsequent transportation to Boot Camp.

MOPP: Mission-Oriented Protective Posture ("MOPP") is a set of layered Personal Protective Equipment (PPE) used by the U.S. military for operations in Nuclear, Biological, Chemical (NBC) environments.

PT: Military jargon for physical exercise: "Physical Training."

PTSD: Post-Traumatic Stress Disorder. A medical condition stemming from exposures to actual or threatened death, serious injury, or sexual violence—including exposures through the experiences of others—that result in persistent negative alterations of an individual's thinking, emotions, sleep, and other behaviors. More fully and formally described in the American Psychological Association's *Diagnostic and Statistical Manual of Mental Disorders (DSM)*.

SEAL: Navy special forces operators. Stands for "SEa, Air, and Land."

S&M: "Sadism & Masochism"

Traumatic brain injury (TBI): A type of intercranial injury often caused by physical blow or penetration, such as from a blast-wave or bullet. Causes can also include falls, collisions, and other violence.

UXO: "UneXploded Ordnance." Bombs, rockets, and other munitions that have been fired or dropped, but have failed to explode.

VA: The U.S. Department of Veterans Affairs.

VAMC: "Veterans Health Administration Medical Center." According to the U.S. Department of Veterans Affairs: "Veterans Health Administration (VHA) is the largest integrated health care system in the United States, providing care at 1,298 health care facilities, including 171 VA Medical Centers and 1,113 outpatient sites of care of varying complexity (VHA outpatient clinics) to over 9 million Veterans enrolled in the VA health care program."

About the Writer

Ron Riekki is a passionate and lifelong explorer of the Great Lakes literary landscape; a prolific creator of poetry, fiction, memoir, theater, and film; and an Operation Desert Shield/Desert Storm veteran of the U.S. Navy. He later served in the U.S. Air Force Reserve. He has received a Pushcart Prize, Shenandoah Fiction Prize, and a selection for The Best Small Fictions.

"Riekki is very important as a regionalist writer, but also as a regionalist editor and promoter for the writings of Northern Michigan and the Upper Peninsula," *American Book Review's* Jeffrey A. Sartain has said. "Because of that, he's very interested in the role writing has and how that relates to notions of place, space, time and context."

In 2019, Riekki authored *My Ancestors are Reindeer Herders and I Am Melting In Extinction* (Apprentice House Press), a collection of non-fiction, short stories, and poetry about the Karelian- and Sámi-American experience. He also wrote *Posttraumatic: A Memoir* (Hoot n Waddle).

Riekki is editor of 2013's *The Way North* (Wayne State University Press), a literary anthology centered on the Upper Peninsula. He is also co-editor of the poetry anthology *Undocumented: Great Lakes Poets Laureate on Social Justice*, and the literary anthologies *Here: Women Writing on Michigan's Upper Peninsula*, and *And Here: 100 Years of Upper Peninsula Writing, 1917–2017*. The latter titles were published by Michigan State University Press.

His 2008 novel, *U.P.,* published by Ghost Road Press, explores the lives of four teenage boys growing up in the Northern Michigan mining towns of Ishpeming and Negaunee. The book was later optioned as screenplay. Riekki has been involved in filming as screenwriter, actor, and producer. He has also co-edited anthologies regarding genre films, such as *The Evil Dead*, *The Twilight Zone*, and Stephen King's *It*.

A former Emergency Medical Technician (EMT), Riekki is currently studying toward a graduate degree in integrated health, mental health, and substance abuse.

Did You Enjoy This Book?

Tell your friends and family about it! Post your thoughts via social media sites, like Facebook, Instagram, and Twitter!

You can also share a quick review on websites for other readers, such as Goodreads.com. Or offer a few of your impressions on bookseller websites, such as Amazon.com and BarnesandNoble.com!

Recommend the title to your favorite local library, poetry society or book club, museum gift store, or independent bookstore!

There is nothing more powerful in business of publishing than a shared review or recommendation from a friend.

We appreciate your support! We'll continue to look for new stories and voices to share with our readers. Keep in touch!

You can write us at:

Middle West Press LLC
P.O. Box 1153
Johnston, Iowa 50131-9420

Or visit: www.middlewestpress.com

www.ingramcontent.com/pod-product-compliance
Lightning Source LLC
Chambersburg PA
CBHW062017040426
42447CB00010B/2031